Beautiful Black Holes
For Kids

Nature Books for Kids
By K. Bennett
Mendon Cottage Books

JD-Biz Publishing

Purchase at Amazon.com

Table of Contents

Introduction

*Space, the final frontier... to explore strange new worlds, to seek out new life, and new civilizations, to boldly go where no man has gone before. ~ **Gene Roddenberry***

***

The universe is full of surprises!

We can find amazing things like galaxies, planets, comets, asteroids, moons, meteorites, and more!

One of the strangest objects we can find in space is called a… **black hole**.

Have you ever heard of black holes? What do you know about them?

Let's learn more!

Black holes are dark areas in space with strong gravity. Not all black holes are black and we cannot see them, but we know they are there.

How do we know they exist even though we can't see them? Scientists study the things that happen around a black hole, and that tells them a black hole is there.

The force of a black hole is so strong light cannot escape. Do you know what happens to light when it gets near a black hole? Strong gravity pulls light and everything else into the center. It is so strong that nothing escapes the powerful force, and everything falls in!

Black holes come in lots of different sizes. Some are big, and some are small. Some black holes are so big; they are called supermassive black holes. That's a big, big hole!

Black holes affect not only space but time too. How so?

Did you know time changes when you get near a black hole? Yes, it does! This is because of Einstein's theory of relativity.

Let's find out how black holes work and what else we can learn about this mysterious force in the universe!

Chapter 1: How do Black Holes Work?

First, we need to learn about Stars.

Stars are bright balls of hot gas that burn in the sky. They have lots of energy and the gas is called plasma. Something special happens inside a star to make it burn so bright. It is called "nuclear fusion."

Nuclear fusion happens when the nucleus of an atom splits and joins with another. This process is powerful and dangerous.

Stars can die. This happens when stars use every bit of its energy and explodes. This explosion is called a supernova.

What is a supernova?

The core of the star must change for a supernova to happen. NASA says we have two types of supernovas:

The first is a binary star system. Binary means two stars. They orbit each other like good friends! One is called a white dwarf. This star is a sneaky little ball. It steals matter from the other stars.

But when it gets too much energy, the white dwarf has nowhere to put it and can't hold it anymore. Can you guess what happens next?

That's right! It explodes into a supernova.

The second is when a star grows old. After it gets too old it's ready to die. The fuel the star needs to keep alive runs out, like a car with no gas.

When the fuel is gone, the mass of the star leaks into the core. When it gets too heavy to hold the mass, the core collapses. But the star doesn't die quietly. It goes out with a big, loud bang! Then a giant explosion rips through space and a supernova is born!

How do supernovas form black holes?

Many stars are big and heavy. And if the star that explodes is heavy enough, a black hole will form.

How black holes are born!

Gravity is an amazing force that keeps things from floating off into space. When a star collapses, gravity pulls on the gas. This causes the star to get smaller and smaller and smaller. After it gets too small, a gigantic amount of mass is pushed into a teeny tiny place. This causes a black hole to form.

Once the black hole is formed, it grows and grows. It can get bigger and bigger until it forms a supermassive black hole.

Not all black holes are supermassive. Some are much smaller. Scientists call these micro black holes.

If you are not sure how a supernova works to form a black hole, here is a simple idea to help you.

Think of planet earth. How big is our home? *Space.com* says that from pole to pole the earth is 7, 900 miles. For our planet to create a black hole, the mass needs to get smaller and smaller until it can fit into the palm of your hand!

How big are black holes?

Black holes come in many sizes.

Ultra-massive black holes:

These black holes are new. Scientists didn't even know they existed. Can you guess how many suns it would take to fit into an Ultra-massive black hole? It's a whopping 10 – 40 billion times the mass of our sun!

Supermassive black holes:

Supermassive black holes have been in space for many years. They are big and take up lots of space. They can be as big as 4 million to 6 billion times the mass of our sun!

Intermediate mass – black holes:

Scientists are not sure if these holes exist, but if they exist, they would be 100 to 1 million times the mass of our sun.

Stellar black holes:

These black holes are smaller than intermediate mass black holes. They are between 5 to tens solar masses big.

It is good to remember that the size of these celestial objects keeps changing. And as scientists learn more and more, they find even bigger black holes in the universe. Who knows what other fascinating things we will learn… soon?

FUN BLACK HOLE FACTS FOR KIDS:

Something incredible happens at the edge of a black hole. It's called an *event horizon*. This part of the black hole is where light disappeared and everything falls into the middle.

If you got as far as the event horizon you might get trapped there if your ship is not strong enough to break free!

Scientists call it the "point of no return."

The gravitational pull is strong in this part of space. This leads to *time dilation.* This means time slows down and stops!

How is this possible?

One of the neat things we know about black holes is that they bend space and time.

That's what causes time to slow down and stop… after you cross the event horizon.

Isn't that amazing?

Chapter 2: Parts of a Black Hole

Let's look inside the different parts of a black hole:

Outer event horizon:

This part of the black hole has strong gravity but not as strong as the event horizon. If you fell into this part of the black hole, you could still break free. But if you keep falling more and more, you will end up in the event horizon.

Event horizon:

Imagine a large bubble or balloon around the black hole. Once you enter this bubble, you cannot escape the pull of the black hole. It would hold you in powerful bands unless you had enough energy to break free.

Imagine this:

One day, you fly out for a friendly visit to the black hole. When you get there, you come out of your ship and drift over the hole. Suddenly, you fall into the opening. If you fall feet first, your legs will stretch and look like strings of spaghetti or chewing gum.

Astronomers have a word for this. Can you guess which one is correct?

1-Calcification
2-Sphagettification
3-Twistification

The correct answer is 2: Sphagettification.

That's a big word! The noodle effect might be easier to remember but it means the same thing. Have you seen what noodles look like?

Long, thin, and stretched out. That is what will happen if you fall into a black hole.

Singularity:

This is another big word but **kidsastronomy.com** has an easier way to say it: a squashed up star! The center of the black hole is called a

---

singularity and gravity is super strong here. This is because lots and lots of mass is squashed into a teeny tiny place.

Anything that falls into the black hole is ripped apart!

DID YOU KNOW?

Light travels fast! It is the fastest speed we know in the universe. If it has nothing to slow it down or stop its speed, light travels at 186,282 miles per second! If you prefer kilometers it's 300,000 thousand kilometers per second.

Can humans go that fast? The answer is… no! We have cars that can go over 250 miles per hour, but it's not as fast as the speed of light. Trains can go over 360 miles per hour but it's still less than the speed of light.

Spacecraft's can go much faster. NASA's New Horizons launched into space at a speed of 45 kilometers per second or 100,000 miles per hour. That's amazing, but it's still less than the speed of light.

Remember: Black holes' trap light even though it's the fastest speed we know. What do you think might happen to a car, train or spacecraft if it falls into a black hole? Can it escape?

Come up with your own ideas and don't forget to share with others! Sharing is a wonderful way to make friends!

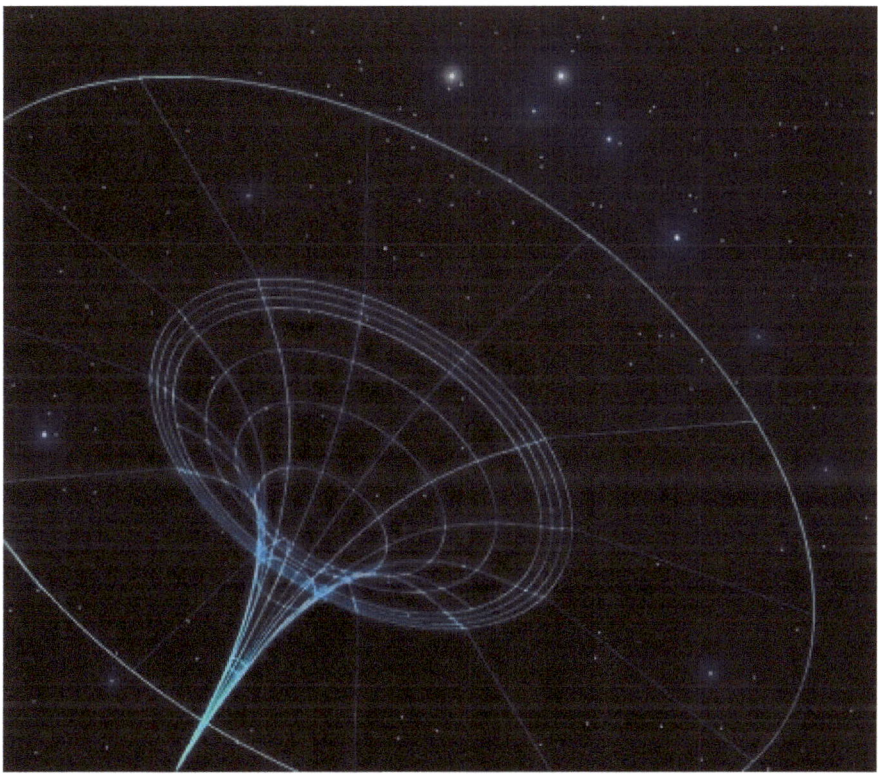

Let's make a black hole!

Experiments are fun and exciting. Let's try to make a model of a black hole and see what other neat things we can learn. Are you ready? These are the supplies you will need:

1- Elastic bandage. Should be light and flexible like the ones used for muscular injuries. (Tubifix).

2- Marble (small)

3- Ball (very heavy)

*Begin by cutting the elastic bandage and make a 40 x 40 centimeters square.

*Stretch it out until it's flat. This is the two dimensions of space. Try to stay as still as possible so the experiment can work! Don't forget to make sure the material stays flat.

*Place the marble on one end. It should roll across to the other side in a straight line. This will make you remember how light travels from one end to the other.

*Place the heavy ball on the bandage. You will see how the ball weighs down the material. It "deforms" space and time. Instead of a straight, horizontal line, the "space" of the bandage will curve around the heavy ball.

*Roll the marble across the bandage. Can it still travel in a straight line?

*This is what happens when a large object bends space and time. It affects everything around it!

*You can try rolling the marble at different speeds to see what it does. Remember: The heavier your object is, the more "deformed" the space will be.

*Soon the marble cannot escape and will simple fall next to the large ball or into the black hole!

Tips: Try a few marbles to see what happens! Then ask yourself these questions: What happens when your marble rolls slower? Does it affect where it falls? What happens if your marble is heavier? What if your ball is heavier? Does anything change?

Let's try another simple experiment.

You will only need two things. First, get your imagination. Yes, the one in your head! Second, get a thimble and put it on the table. Ready? Great!

Put your books in it. (**Remember:** Use your IMAGINATION). Don't stuff your books into a thimble for real. (They can't fit anyway!)

After you add your books, add your clothes, shoes, teddy bears, and bed. Then take the household furniture and cram them into the opening. Add the floor, the roof, the walls, and the foundation. Keep adding until you add a whole earth's full! Good job!

Now, let's get more creative. Think of something much bigger. Imagine the state of New York. If a black hole was as big as the state, it would have the mass and gravity of the sun. All the planets in our solar system would fit inside this black hole. Isn't that amazing?

Chapter 3: Interesting Facts

I hope you are enjoying this book on Beautiful Black Holes! Here are a few more neat facts:

- The sun is a big star, but it does not have enough mass to become a black hole! Isn't that good news?

-There are lots of black holes in the universe. We do not the exact number but scientists predict a black hole is born every day! That's lots of black holes.

- Scientists say there's a large black hole in the center of our Milky Way Galaxy. It's very far away. 30,000 light years! And it's 30 million times the size of our sun!

-Black holes make lots of noise. It is not the usual sounds we hear, but NASA detected "sound waves" coming from a black hole. It is approximately 250 million light years away and super massive!

-Time slows down in a black hole. Even a minute near the event horizon can be years on earth! This means you will age much slower than you would on earth.

-Black holes don't look like a funnel. They look more like a sphere, and they spin.

-Black holes don't last forever. After a while they evaporate into space!

-Stephen Hawking said in 1974 that black holes take in matter, as well as spit it out too!

-Scientists continue to add new information to their theories on black holes. There is lots they don't know but they're still searching!

FUN QUIZ FOR KIDS:

Quizzes can be a fun and exciting way to learn. This quiz is adapted from **Science.com.nz** and teaches us fascinating things about the solar system.

Put on your thinking cap and have fun!

1 – Name the closest planet to the sun.

2 – Can you name the second biggest planet in our solar system?

3 – Can you guess the name of the hottest planet in our solar system?

4 – What planet is famous for a large, red, spot?

5 – Which planet has beautiful rings?

6 – Can humans breathe in space?

7 – What is our sun… a star or a planet?

8 – Can you remember the name of the first person to walk on the moon?

9 – Which planet is called the red planet?

10 – What force is holding us to the earth? Hint: It is the same force that keeps us from floating off into space.

11 – Are human beings on Mars? Has anyone ever gone there?

12 – Which place uses telescopes and scientific equipment to study the wonders of space and astronomy?

13 – Do you remember the name of NASA's most famous telescope?

14 – Earth is found in which galaxy?

15 – What is the name of the first satellite sent into space?

16 – Ganymede is a moon that belongs to which planet?

17 – What is the name of Saturn's biggest moon?

18 – Where can you find the largest volcano in the solar system? It's called Olympus Mons.

19 - Does the sun go around the earth?

20 – Is Neptune bigger than planet earth?

Finds the answers after the conclusion!

Vocabulary:

Our universe is full of amazing things. Here is a small list of vocabulary words to help you learn more.

-White dwarf   -Dark matter

-Black dwarf   -Remote sensing

-Constellation   -Interstellar

-Force   -Pulsar

-Embryonic star cloud -Quasar

-Cosmic   -Spectroscope

-Epic   -Nebula

-Cosmos   -Phenomenon

-Pulsar

-Quasar

-Sidereal

-Singularity

-Umbra

-Speed of light

-Wavelength

Do you know what these words mean? If you are not sure, ask your parent or a guardian's permission to search for the definition. I hope you learn something new!

(www.dictionary.com)

Conclusion:

***In conclusion*:**

Our universe is full of wonderful things and black holes are an amazing part of it. They are an enigma, which means a puzzle. We don't know a lot about them, but we continue to learn more and more. Imagine what we will learn one day!

Einstein's theory of relativity is used to explain what happens at a black hole. He discovered Newton's law of gravity did not work when objects were huge and far away.

He changed the law to become the theory of relativity, which says if you want to know if you are moving you must be looking at another object. By doing this you can see if you are getting further away, closer, or left or right of the object, showing that you are moving.

Albert Einstein also proved that light moves at the same speed no matter where it is or what is happening to it.

Lastly, he concluded that large objects can cause outer space to bend, like your mattress when you climb into bed.

Something else to think about!

Black holes are a great choice for a science experiment. If you use this topic, don't forget the steps you need to make it a great science project.

Steps:

1 – You need to ask a **question** to be answered by observation or experimentation. Make it a very interesting question so your classmates and teachers will want to learn the answers!

For example: What's inside a black hole? What would it feel like if you could go inside the core? Do you think you'll be able to travel to another dimension? What would it look like? Will it differ from planet earth? In what ways might it be different?

2 – The next step is to state a **Hypothesis**. This is a big word but Sciencekidsathome.com explains it like this*: It is a tentative explanation for an observation, phenomenon, or scientific problem that can be tested by further investigation.*

Your hypothesis is what you think the results of your project will be when your research is all done!

Write your ideas on what you think you might discover on black holes.

3 – Next on the list is: **Procedure.** This is very important. Procedure will help you discover the answer to your question and prove what you are trying to say.

There are other experiments online that can help you. Ask your parent or a guardian to help you search. Or ask for permission before you search.

4 – **Results**. You will need to show your results and all the information you collected for your project.

5 – **Conclusion**. Finish up with what you learned and then answer the question you had in Step 1. If you can't answer the question, explain why the question cannot be answered.

I know you will have fun learning about black holes and all its wonders!

If you don't like the ideas in this book, put on your thinking cap and come up with your own conclusions! I am sure you will do an amazing job!

We hope you have enjoyed this book on Beautiful Black Holes. Always remember…

"Educating the mind without educating the heart is no education at all." - *Aristotle*

Happy Learning!

Space Quiz Answers

1. Mercury

2. Saturn

3. Venus

4. Jupiter

5. Saturn

6. No

7. A star

8. Neil Armstrong

9. Mars

10. Gravity

11. No

12. An observatory

13. Hubble Space Telescope

14. The Milky Way Galaxy

15. Sputnik

16. Jupiter

17. Titan

18. Mars

19. No

20. Yes

Sources:

http://www.kidsastronomy.com/

http://www.physlink.com/Education/AskExperts/ae253.cfm

http://www.nasa.gov/audience/forstudents/5-8/features/nasa-knows/what-is-a-black-hole-58.html

https://student.societyforscience.org/article/black-hole-mysteries

http://www.freewordsearches.net/wordsearch/black-holes-in-space1

http://www.unawe.org/activity/eu-unawe1308/

https://kidskonnect.com/science/black-holes/

http://www.kidsastronomy.com/academy/lesson210_assignment3_8.htm

Author Bio

K. Bennett loves to write for both children and adults. Many subjects are interesting to research, but writing for children is special to her heart.

Her favorite pastimes include reading, traveling and discovering new things. Each of these activities helps to fuel her imagination and acts like a blank canvas waiting for more stories.

She is intrigued with fantasy elements like hidden worlds and faraway lands. And basically anything that gets her imagination soaring to new heights!

Her writing credits include children books online, short stories for online magazines, and novellas listed at Amazon.com

Download Free Books!

http://MendonCottageBooks.com

Our books are available at

1. Amazon.com

2. Barnes and Noble

3. Itunes

4. Kobo

5. Smashwords

6. Google Play Books

Download Free Books!
http://MendonCottageBooks.com

Publisher

JD-Biz Corp

P O Box 374

Mendon, Utah 84325

http://www.jd-biz.com/

Mendon Cottage Books

P O Box 374, Mendon Utah 84325

www.ingramcontent.com/pod-product-compliance
Lightning Source LLC
Chambersburg PA
CBHW041116180526
45172CB00001B/283